HOW TO CHOOSE THE RIGHT PLACE TO LIVE

FEW THINGS YOU MUST KNOW BEFORE

ENTERING A NEW NEIGHBORHOOD

A.D RAMS

Contents

CHAPTER ONE 3

INTRODUCTION 3

The significance of choosing the ideal place to call home 9

Determining Individual Preferences for Lifestyle and Priorities15

CHAPTER TWO 20

Evaluating Cost of Living and Affordability 21

Assessing Employment Market and Professional Prospects.....28

Investigating School Systems and Education 34

Analyzing Medical Services and Facilities 41

CHAPTER THREE 43

Taking Crime and Safety Into Account.....................47

Examining Options for Commuting and Transportation 54

CHAPTER FOUR 61

Analyzing Neighborhood and Community Features 61

Investigating Possible Sites and Compiling Data.....................68

CHAPTER FIVE 72

Choosing a course of action and moving forward....................74

Summary 80

THE END 85

CHAPTER ONE

INTRODUCTION

Selecting where to live is an important choice that can have a big influence on your opportunities, happiness, and overall quality of life. Several things should influence your decision-making process, regardless of whether you're moving to a new town, inside your existing city, or even abroad. This overview will assist you in navigating the challenges of choosing the ideal residence.

Establish Your Priorities: Spend some time deciding what is most important to you before starting your hunt for a new residence. Take into account elements including the cost of living, the weather, employment prospects, the distance to family and friends, cultural attractions, leisure pursuits, and availability to healthcare. You can focus on places that fit your lifestyle preferences and reduce your alternatives by making your priorities clear.

Investigate Possible places: After deciding on your top priorities, look into possible places that fit your requirements. To learn more about various regions, use online resources like forums, real estate websites, and city guides. Be mindful of things like the cost of housing, the level of crime, the caliber of the schools, the availability of transit, and the facilities in the area. Visit potential communities as well to have a sense of the neighborhood's vibe and sense of community.

Assess Economic Opportunities: A location's suitability for long-term habitation is largely dependent on its economic prospects. Examine the employment landscape in the area of your choice, taking into account industry diversity, employment growth trends, and median salary levels. Take into account other elements as well, such as the existence of significant employers, chances for entrepreneurship, and the region's general economic stability.

Take Lifestyle Aspects Into Account: Your dream home should complement your interests and lifestyle choices. If you're an outdoor enthusiast, give priority to places with lots of greenery, parks, and leisure centers. If you enjoy culture, look for areas that are well-known for their thriving theaters, museums, and food scenes. Assessing lifestyle elements guarantees that you will have contentment and involvement in your new community.

Evaluate Housing Options: When deciding where to reside, availability and affordability of housing are important factors. To gain insight into housing trends, property valuations, rental rates, and housing inventory, conduct research on the local real estate market. Consider things like the kind of housing (apartment, single-family home, condominium), the features of the community, and the facilities in the housing. When assessing housing possibilities, don't forget to account for your budget and long-term housing objectives.

Think About Future Growth and Development: Planning ahead for future growth and development might help you choose a home with greater knowledge. Examine the location you have selected's demographic changes, infrastructure developments, and urban planning efforts. Think about the potential effects these variables may have on quality of life, property values, and general community dynamics in the future.

Seek Community Connection: Having a sense of community and belonging is crucial to your general wellbeing. Look for ways to interact with locals, become a member of the community, and take part in social events. In addition, take into account the accessibility to facilities such as volunteer opportunities, community centers, libraries, and places of worship that share your values and interests.

Consult Local Experts: When in doubt, get advice from local professionals like real estate brokers, moving consultants, or local authorities. Because they are familiar with the local real estate market, community dynamics, and quality of life concerns, these people may offer insightful advice.

You can choose the ideal place to live that fits with your priorities, tastes, and long-term objectives by carefully weighing these variables and doing extensive study. Keep in mind that while settling on the ideal location could necessitate some compromise, putting your priorities first will ultimately result in a happy living situation.

The significance of choosing the ideal place to call home

Choosing a place to call home is crucial since it affects so many different facets of a person's life. The following are some of the factors that make location selection so important:

Quality of Life: Your overall quality of life is greatly impacted by the location of your home. Access to amenities, leisure activities, medical facilities, and green areas are a few examples of the kinds of factors that affect your happiness and well-being in your current residence.

Career Opportunities: The labor market and career opportunities differ by region. Your chances of landing a good job and developing in your career are increased when you choose a place with a strong economy and a variety of industries.

Cost of Living: Living expenses differ significantly between locations. Your capacity to maintain financial stability can be severely impacted by daily spending, taxes, housing rates, and energy bills. A comfortable living and financial security can be attained by selecting an inexpensive place that fits your budget.

Education System: The quality of the education system is a crucial factor for those who are planning a family or who already have children. A top-notch education and a solid foundation for future success are certain when you choose a place with respectable schools, institutions, and educational resources for your kids.

Community and Social relationships: Your general happiness and well-being are influenced by the sense of community and social relationships that are established in a certain region. Meaningful relationships and a sense of belonging are facilitated by living in a place with a strong sense of community, a wide range of social opportunities, and shared values.

Health and Safety: When deciding where to reside, it's crucial to take into account the area's healthcare system and safety standards. Your physical and mental health are influenced by a variety of factors, including environmental conditions, crime rates, emergency response times, and access to high-quality healthcare services.

Transportation and Accessibility: Your everyday life is greatly impacted by the convenience of getting to and from work, important services, and recreational areas. Selecting a place that has easy access to major highways, good public transportation, and well-kept roads guarantees convenient travel and increases your mobility overall.

Environmental Factors: Your health and general quality of life can be greatly impacted by environmental factors, which include climate, air quality, natural catastrophes, and closeness to environmental risks. A healthier and more pleasurable living environment is encouraged by choosing a site with a good climate and few environmental hazards.

Long-Term Investment: Selecting a suitable place to call home is another long-term investment. The possible return on investment for real estate investors and homeowners can be influenced by future development plans, property valuations, and economic stability.

Personal Fulfillment: In the end, deciding where to live comes down to finding a place that fits with your long-term objectives, values, and preferences. A happy and rewarding life experience is facilitated by residing in an area that advances your career and personal goals, cultivates deep connections, and improves your general well-being.

It is impossible to overestimate the significance of choosing the ideal area to call home. The location of your home has a significant impact on your quality of life and future prospects, affecting everything from your social connections and general happiness to your professional options and financial security. In order to discover a place that suits your needs and improves your general well-being, much thought and research are necessary when deciding where to reside.

Determining Individual Preferences for Lifestyle and Priorities

It's critical to determine your own priorities and lifestyle choices before settling on a residence. You will be able to make an informed choice that is in line with your needs, values, and goals with the aid of this self-reflection process. The following can help you determine your own priorities and lifestyle choices:

Assess Your demands: To begin, assess your fundamental demands and necessities for day-to-day existence. Think on things like how big of a place you need to stay, how many bedrooms and bathrooms you require, and if you want an urban, suburban, or rural environment. If you have dependents, evaluate their particular needs in terms of healthcare, education, and accessibility.

Think About Your Values: Consider your values and the things that are most important to you in life. Do you think that professional possibilities and career progression are important? Do you think having strong social ties and a feeling of community is important? Do you find outdoor recreation and environmental sustainability to be important? Prioritizing things that fit with your lifestyle and beliefs will be made easier if you are aware of your basic principles.

Think About Career and Economic Opportunities: Assess your professional aspirations as well as the economic prospects in various regions. Would you be open to moving in search of a better job, or would you rather put your lifestyle above your career? When evaluating the economic prospects in possible places, take into account elements like the cost of living, industry diversification, job availability, and income potential.

Examine Cultural and Recreational Amenities: When assessing cultural and recreational amenities in possible locales, take your hobbies and interests into account. Do you like going to restaurants, seeing live shows, or visiting galleries and museums? Do you like hiking, biking, or water sports as an outdoor hobby? Evaluating the accessibility of recreational opportunities, cultural attractions, and entertainment options will assist you in selecting a place that complements your preferred way of life.

Think About Social ties and Community: Give some thought to the significance of social ties and community in your life. Are you trying to find a close-knit community where people go by names? Which would you prefer—access to a vast array of social possibilities or a diversified, global urban environment? Think about your social preferences and how different places could satisfy your demand for social interaction and community.

Examine Healthcare and Education Facilities: If you have dependents or children, consider the facilities' quality when considering different places for your healthcare and education needs. Do your homework on school districts, curriculum, and extracurricular activities to make sure your kids get a great education. Additionally, make sure your healthcare needs are satisfied in the place of your choice by evaluating emergency services, experts, and healthcare facilities.

Think About Environmental elements: Your quality of life can be greatly impacted by environmental elements including the weather, the quality of the air, and the surrounding environment.

CHAPTER TWO

Think about whether you would rather live in a warmer place with constant sunshine or a milder place with four distinct seasons. When selecting a location, consider environmental concerns such pollution, natural catastrophes, and close proximity to environmental hazards.

Think About Your Long-Term Objectives: When selecting a residence, take your long-term objectives and desires into account. Are you trying to find a place to call home and build a family? Do you see yourself going back to school, launching a company, or living in a certain place when you retire? You can select a place that supports your future aims and offers chances for both professional and personal improvement by thinking back on your long-term objectives.

Finding a place to live that fits your requirements, values, and goals can be accomplished by carefully considering your unique objectives and lifestyle choices. Think about what is most important to you, look into possible places, and evaluate how various aspects will affect your general well-being and quality of life.

Evaluating Cost of Living and Affordability

When selecting a place to reside, it is important to consider affordability and the cost of living as these factors have a direct impact on your entire quality of life and financial stability. Here's how to assess cost of living and affordability effectively:

Establish Your Budget: To begin, figure out how much you can afford to pay each month for housing, utilities, groceries, transportation, medical care, and other expenses. When creating your budget, take into account your income, savings, debt commitments, and financial objectives.

Investigate Housing expenses: To determine whether various housing options are affordable, investigate housing expenses in prospective locales. Examine the median cost of homes, rental pricing, property taxes, and premiums for homeowners' insurance in different cities or neighborhoods. When analyzing housing costs, take other aspects like the size of the home, the amenities, and its accessibility to work hubs into account.

Examine Utility Costs: Determine the costs associated with utilities in possible locations, including internet, cable or satellite TV, water, gas, and electricity. To accurately estimate your monthly utility prices, look up the average utility costs for homes with similar size and usage patterns.

Think About Transportation expenditures: Depending on where you live and how you commute, your expenditures can vary greatly. Consider the cost and accessibility of public transportation choices, including buses, trains, and subways, in addition to the expenses associated with owning and operating a car, such as gas, insurance, maintenance, and parking. To save on transportation costs, pick a place with accessible and reasonably priced transit choices.

Compare Grocery and Consumer Prices: In order to determine the cost of living in a potential place, research the prices of groceries and consumer products. To determine if daily expenses are affordable, compare the costs of necessities like food, clothing, housekeeping supplies, and personal care goods. When assessing consumer prices, take into account variables like the existence of wholesale clubs or discount stores, as well as sales tax rates.

Evaluate Healthcare prices: A number of variables, including insurance coverage, healthcare providers, and local trends in healthcare pricing, can affect healthcare prices. Examine whether medical services, doctors, experts, and healthcare facilities are offered in possible areas. To precisely estimate your medical expenses, consider your health insurance alternatives, premiums, deductibles, and out-of-pocket spending.

Factor in Other Expenses: When assessing the total cost of living in possible places, take into account additional costs like childcare, education, entertainment, eating out, and savings contributions. To determine whether lifestyle costs are affordable, check into the availability and cost of daycare centers, educational institutions, theaters, leisure centers, and cultural attractions.

Compute the Cost of Living Index: To compare the affordability of various regions, use cost of living indices or calculators. These tools provide an overall cost of living comparison between cities or areas by taking into account several elements like housing costs, transportation expenses, food prices, healthcare costs, and other living expenses. In order to get a more precise estimate of affordability, think about utilizing many cost of living calculators.

Determine Affordability Based on Income: Determine how affordable housing and living expenses are in relation to your income level to make sure they stay within your means. To prevent financial hardship, aim for a home cost-to-income ratio and a total debt-to-income ratio that are within advised bounds.

Examine Long-Term Financial Effects: Evaluate the long-term financial effects of a certain place, taking into account things like potential future home appreciation, employment prospects, possibility for income growth, and retirement savings. Select a place where you can reach your financial objectives and sustain long-term financial security.

You can choose a place to live that fits your budget, financial objectives, and general quality of life by carefully evaluating affordability and the cost of living in various areas. To make sure you can easily afford your selected place, make sure you take into account all pertinent elements and investigate the local economy.

Assessing Employment Market and Professional Prospects

When deciding where to reside, it's important to consider the job market and career chances because these factors have an immediate impact on your employment prospects, earning potential, and ability to grow in your career as a whole. Here's how to evaluate possible locations' employment markets and career opportunities:

Investigate Local Industries: To begin, find out which sectors and industries are most prevalent in possible areas. Determine the important employers, big businesses, and industry clusters that promote job growth and the economy. Take into account elements like the diversity of industries, growth patterns, and the need for skilled labor across many industries.

Examine Job Availability: Determine whether positions and employment prospects are available in your desired field or industry. To find job openings and vacancies in possible places, use professional networking sites, corporate websites, and online job boards. When assessing job availability, take into account the quantity of job opportunities, jobs growth forecasts, and rivalry for employment.

Determine Your Income Potential: To determine your income potential, look into the median income, salary ranges, and potential for income growth in prospective regions. When assessing income potential, take into account variables including industry-specific remuneration norms, cost of living adjustments, and regional salary trends. Make sure your pay is commensurate with the cost of living and the state of the local economy by comparing income levels in other places.

Think About Industry Specialization: Determine whether possible sites have a focus or specialization in sectors related to your field of work or professional objectives. To optimize job chances and development prospects, pick a place with a strong presence in industries that match your experience, talents, and career goals. When evaluating industry specialization, take into account elements including availability of professional development resources, networking opportunities, and industry associations.

Investigate Networking Opportunities: To grow your professional network and gain access to career resources, investigate professional communities and networking opportunities in prospective locales. To network with area professionals, employers, and recruiters, go to industry events, conferences, seminars, and networking mixers. Join online forums, alumni associations, and professional associations to have access to job leads, mentorship possibilities, and career advice in your preferred area.

Examine Educational and Training Materials: Determine whether educational and training materials are offered in possible places so that you can improve your credentials and abilities for professional growth. Look into the relevant courses, certificates, and training possibilities offered by colleges, universities, vocational institutions, and continuing education programs. When assessing educational resources, take into account elements including program reputation, faculty expertise, and accreditation.

Examine Telecommuting and Commuting choices: Determine whether it is feasible to work remotely or commute to work by analyzing telecommuting policies and choices in prospective locations. To evaluate the price and convenience of traveling to job centers, research parking availability, traffic congestion, and public transportation choices. Examine whether your employer has opportunities for telecommuting, remote work, or flexible work schedules that fit your tastes and way of life.

Analyze Career Growth chances: To guarantee long-term career happiness and professional development, assess career growth chances and progression prospects in prospective locales. Examine variables including leadership chances, work mobility, promotion rates, and availability of career coaching and mentoring programs. Pick a place where you may advance your career, improve your skills, and move up the corporate ladder in the sector or profession of your choice.

You can choose where to reside that best suits your professional aspirations, expected income, and chances for overall career progress by carefully assessing the employment market and career options in possible regions. To ensure that your choice is the greatest one for your career success and fulfillment, make sure you take into account all pertinent criteria, do extensive research, and consult with mentors, career counselors, and industry professionals.

Investigating School Systems and Education

When making a relocation decision, particularly if you have children or intend to establish a family, it is imperative that you do your homework on education and school systems.

Here's how to assess educational opportunities and school systems in possible locations:

Investigate School Districts: Begin by looking into the school districts in possible places. Seek out details on school zones, district demographics, and boundaries. Determine whether high-achieving school districts have high graduation, college admittance, and academic success rates.

Examine school ratings and rankings from reliable sources, including local publications, educational websites, and the Department of Education. When assessing school rankings, take into account elements including test results, student-to-teacher ratios, teacher qualifications, and extracurricular programs offered.

Examine School Types: Take into account the public, private, charter, magnet, and foreign school options that may be found in your prospective location. Examine the curricula, instructional strategies, academic offerings, and special education resources provided by various school types to ascertain which one best suits your child's learning requirements and preferences.

Investigate Specific Schools: Look at specific schools in the districts where you plan to teach or institutes of higher learning. To find out more about the culture, values, and academic offers of each school, visit their websites, read parent reviews, and arrange for tours or informational sessions with staff members and administrators.

Evaluate Special Education Services: Find out where special education resources and services are offered if your child needs more help or has special education needs. To make sure your child gets the right accommodations and services, consider the caliber of the special education programs, individualized education plans (IEPs), and support services provided by school districts or educational institutions.

Think About Extracurricular Activities: Assess the availability of clubs, sports programs, extracurricular activities, and enrichment courses that schools in prospective areas offer. When evaluating extracurricular activities, take into account elements like participation rates, range of extracurricular offers, and chances for leadership development and student engagement.

Investigate School Resources and Facilities: Look into the amenities, resources, and facilities offered by the schools in possible locations. Examine elements like athletic facilities, libraries, performing arts venues, school infrastructure, and classroom technology to make sure that schools foster student development and offer a favorable learning environment.

Examine School Safety and Discipline rules: To guarantee a secure and encouraging learning environment for your child, consider the school safety and discipline rules at prospective sites. Examine school safety protocols, emergency action plans, anti-bullying initiatives, and disciplinary guidelines to make sure that educational institutions put students' welfare first and sustain a healthy learning environment.

Think About Parent and Community Involvement: Take into account the degree of community and parent involvement in the schools in your possible locations. To evaluate chances for parent and community involvement in school events, activities, and decision-making processes, investigate parent-teacher organizations (PTOs), school advisory committees, and volunteer possibilities.

Get Input from Locals: Find out what locals, neighbors, and community members have to say about the educational and school systems in possible locales. To learn more about the advantages and disadvantages of the local educational system, ask for advice, opinions, and firsthand accounts regarding the local schools.

You can choose where to live in a way that best meets your child's educational needs, academic achievement, and general well-being by doing extensive study on education and school systems in possible neighborhoods. To make the greatest decision for your family, make sure to take into account all pertinent criteria, personally visit schools, and speak with locals and educational professionals.

Analyzing Medical Services and Facilities

While deciding where to live, it is important to consider healthcare facilities and services because having access to high-quality medical care can have a big impact on your health and quality of life. Here's how to assess medical services and facilities in possible places effectively:

Investigate Healthcare Providers: Begin by looking into probable areas' hospitals, clinics, medical centers, and specialists. Find out what medical specializations are offered, what kinds of healthcare services are provided, and whether or not a facility is accredited or certified.

Review hospital rankings and ratings from reliable sources, such as the U.S. Centers for Medicare & Medicaid Services (CMS), in order to evaluate hospitals. Healthgrades and News & World Report. When assessing hospital ratings, take into account variables including patient outcomes, safety precautions, infection rates, and patient satisfaction levels.

Evaluate Primary Care Access: Evaluate the availability of primary care providers in prospective locations, such as family doctors, internists, pediatricians, and general practitioners. To guarantee prompt access to routine medical treatments and preventative care, look into the availability of urgent care facilities, clinics, and primary care practices.

CHAPTER THREE

Examine specialized Care Services: Examine whether medical specialists and specialized care services are available in possible locations. To guarantee that you have access to specialized medical treatment when you need it, find out who the specialists are in fields like cardiology, cancer, orthopedics, obstetrics and gynecology, and mental health.

Investigate Emergency Care services: Look into suitable locations for emergency rooms, urgent care centers, and trauma centers, among other emergency care services. To guarantee that patients have timely access to high-quality emergency medical treatment, consider variables such emergency response times, wait periods, and the availability of critical care services.

Examine Healthcare Insurance choices: In prospective locations, examine healthcare insurance choices such as government-sponsored healthcare programs, individual health insurance plans, and employer-sponsored health plans. To guarantee that you have access to reasonably priced healthcare coverage, check into the availability of healthcare insurance providers, coverage options, premiums, deductibles, and out-of-pocket expenses.

Evaluate Telehealth Services: Take into account whether virtual healthcare choices and telehealth services are offered in prospective locations. To evaluate access to remote healthcare services and telemedicine technology, conduct research on telemedicine providers, virtual consultations, and online healthcare platforms.

Investigate Medical Facilities Near Your Home: To guarantee easy access to medical services, investigate medical facilities close to your home or a potential neighborhood. When considering the accessibility of healthcare facilities, take into account elements like the distance to pharmacies, medical centers, hospitals, and clinics.

When assessing healthcare facilities and services in possible locations, take into account performance indicators and healthcare quality measurements. To evaluate the overall quality of healthcare services rendered, research measures like patient outcomes, readmission rates, infection rates, and quality improvement activities are used.

Get Input from Locals: Find out what people in the neighborhood, community, and local inhabitants think about the healthcare facilities and services in possible sites. To learn more about the healthcare environment in the area, ask for advice, opinions, and first-hand stories concerning local medical professionals and healthcare providers.

You can choose where to live that emphasizes access to high-quality healthcare and supports your general health and well-being by carefully evaluating healthcare facilities and services in various places. For the best decision regarding your healthcare needs, make sure you take into account all pertinent considerations, investigate your local healthcare options, and speak with medical professionals and locals.

Taking Crime and Safety Into Account

Picking a place to live has a significant impact on your general quality of life, mental health, and general well-being, thus it's important to take safety and crime statistics into account. Here's how to evaluate possible locations' safety and crime rates effectively:

Investigate Crime Statistics: To begin, look up the crime rates in any prospective locations. Seek out statistics on crime rates from reputable sources including regional law enforcement offices, official databases, and crime mapping websites. Pay particular attention to violent crime, property crime, and general crime rates. To gauge the relative safety of several neighborhoods, cities, or regions, compare their crime statistics.

Assess Crime Trends: Over time, assess probable areas' crime trends and patterns. To find any swings or changes in crime rates, criminal hotspots, and crime prevention initiatives, look at past crime data and trend studies. When evaluating crime patterns, take into account variables including population growth, prevailing economic conditions, and tactics used by law enforcement.

Evaluate Neighborhood Safety: Evaluate each prospective location's neighborhood's level of safety. To find safe and secure neighborhoods with low crime rates and less safety concerns, look for neighborhood crime statistics, crime maps, and safety ratings. When evaluating neighborhood safety, take into account elements like home security systems, neighborhood watch initiatives, and community policing initiatives.

Think About Crime Types: Take into account the kinds of crimes that are common in possible places. Analyze the frequency of property crimes like theft, vandalism, and burglary as well as violent crimes like assault, robbery, and homicide. Prioritize safety concerns based on the possibility of experiencing various criminal activities.

Investigate Safety Initiatives: Look at neighborhood policing initiatives, crime prevention plans, and safety measures in prospective locales. Seek out details about safety awareness campaigns, workshops on crime prevention, neighborhood watch programs, and collaborations between local organizations and police enforcement. Examine how well safety programs are working to lower crime rates and increase community safety.

Consult Crime Victim Resources: For information on safety and crime concerns in possible places, consult advocacy groups, support groups, and crime victim resources. Consult locals, community leaders, and advocates for victims of crime for their perspectives on crime, safety concerns, and local attempts to address safety issues.

Evaluate Law Enforcement Response: Evaluate how well-responsible and efficient local law enforcement organizations are in handling safety and criminal issues in possible places. Examine response times, criminal clearance rates, and the cooperation between law enforcement and community partners. When assessing law enforcement reaction capabilities, take into account the availability of resources such as detectives, patrol officers, and specialized units.

Take Environmental aspects into Account: Take into account environmental aspects that could affect crime and safety rates in possible areas. To determine environmental safety, consider elements like illumination, visibility, street design, public areas, and surrounding landscape. Keep an eye out for infrastructure that is conducive to pedestrian traffic, well-kept public spaces, and initiatives to enhance environmental security and deter crime.

Get Input from Locals: Find out what the locals, neighbors, and community members think about the safety and crime rates in possible places. Request advice, opinions, and firsthand accounts of crimes, safety issues, and initiatives to enhance neighborhood safety. When assessing safety and crime concerns in the area, take the opinions of the locals into consideration.

Visit and tour communities: To get a personal assessment of safety and criminal concerns, visit and tour communities within possible sites. Pay attention to details like building security, street illumination, general neighborhood attractiveness, and cleanliness. During your visits, keep an eye out for indications of social cohesion, community involvement, and active participation in crime prevention initiatives.

You may make a well-informed decision about where to live that puts your safety, security, and peace of mind first by carefully evaluating crime statistics and safety in various areas. Making the best decision for your safety needs requires careful consideration of all pertinent criteria, research into local safety efforts, and consultation with law enforcement and local communities.

Examining Options for Commuting and Transportation

Examining your alternatives for commute and transportation is crucial when deciding where to live because it affects your daily commute, accessibility, and general mobility. Here's how to investigate commuter and transit choices in possible areas efficiently:

Examine Public Transportation: Look into the accessibility and dependability of public transportation choices in possible sites, such as light rail systems, buses, trains, subways, and commuter ferries. Check with your local transit authority, transportation authorities online, and transit apps for information on routes, schedules, fares, and frequency of operation. When assessing public transportation options, take into account the transit system's coverage area, accessible features, and facilities provided.

Evaluate Commute Times: Determine how long it will take you to commute to your job, school, or other destinations, as well as the best routes to take. To predict journey times, traffic conditions, and other routes, use online mapping tools, navigation applications, and commute planners. When evaluating commute times, take into account variables including travel time peaks, congestion, and distance.

Assess Walkability and Bikeability: Determine how walkable and bikeable prospective communities and neighborhoods are. To evaluate how easy it is to walk and bike in the neighborhood, look up the walk scores, bike scores, and pedestrian infrastructure. Seek out shared-use paths, bike lanes, sidewalks, and crosswalks that enhance the safety and connectedness of bicyclists and pedestrians.

Examine Major Highway Access: Take into account how accessible major highways, expressways, and arterial roads are from possible locations. To evaluate accessibility to local attractions and travel hubs, investigate the vicinity of important transportation routes and interchanges. When assessing access to major highways, take into account variables including traffic congestion, road quality, and proximity to important job hubs.

Look into Parking Availability: If you own a car or depend on driving for transportation, look into parking availability and possibilities in possible places. When evaluating parking availability, take into account elements like parking permits, parking costs, off-street parking facilities, and on-street parking. In communities and neighborhoods, look for parking enforcement rules and residential parking regulations.

Examine Ride-Sharing and Carpooling Options: Examine the ride-sharing and carpooling options such as ride-hailing, carpool matching, and vanpooling that are offered in possible areas. To increase the availability of shared transportation options and decrease the number of lone drivers, look into employer-sponsored commuting programs, carpool lanes, and ride-sharing applications.

Think About Telecommuting and Remote Work: As part of your transportation and commute plan, take into account choices for telecommuting and remote work. Examine the telecommuting resources, rules, and flexible work schedules that companies in prospective locations provide. To lessen the necessity for a daily drive, look into alternatives to work from home or telecommute part-time.

Analyze Transportation Costs: Determine the transportation expenses related to various routes of travel in possible destinations. To find the most economical way to commute for your budget, compare the costs of parking, fuel, car maintenance, public transportation fares, and tolls. When estimating transportation costs, take into account elements like tax advantages, employer-sponsored commuter perks, and transit subsidies.

Investigate Upcoming Transportation Projects: Look into upcoming transportation initiatives and planned infrastructure upgrades for possible sites. Seek out details on initiatives involving bike lanes, pedestrian improvements, road renovations, and transit expansions that could increase the area's connectivity and transit options. When assessing possible locations, take into account the possible effects of upcoming transportation projects on property values and commuting patterns.

Get Input from Locals: Find out what locals, neighbors, and community members have to say about the commute and transit alternatives in possible locations. Request advice, opinions, and first-hand accounts of the area's other transit options, commuting techniques, and transportation problems. When assessing the local transportation and commute choices, take locals' opinions into consideration.

You may choose where to live that best suits your mobility, accessibility, and convenience by carefully examining the commute and transit alternatives in possible places.

CHAPTER FOUR

To make the best decision for your transportation needs, make sure to take into account all pertinent considerations, investigate local transit options, and speak with locals and transportation providers.

Analyzing Neighborhood and Community Features

When deciding where to live, it is important to consider the qualities of the community and neighborhood because these factors have an immediate impact on your way of life, sense of belonging, and general well-being. Here's how to evaluate neighborhood and community features in possible locations:

Examine Demographics: Look at the makeup of possible communities and neighborhoods. Seek data on household composition, age distribution, population size, ethnic diversity, and socioeconomic status. When assessing neighborhood demographics, take into account elements like cultural diversity, community cohesion, and demographic trends.

Examine Safety and Crime Rates: Examine possible neighborhoods' safety and crime rates. To determine the frequency of crime and safety concerns in the area, look up crime data, crime maps, and safety ratings. When assessing neighborhood safety, take into account variables like total crime rates, property crime, and violent crime.

Examine School Quality: If you have children or intend to create a family, pay close attention to the quality of the schools in any prospective communities. Examine test results, graduation rates, parent evaluations, and school ratings to assess the academic standing and reputation of nearby schools. When evaluating the quality of a school, take into account elements like extracurricular activities, special education programs, and school amenities.

Examine Amenity and Service Access: Take into account the neighborhoods' respective service and amenity accessibility. To determine convenience and accessibility, consider how close grocery stores, malls, dining establishments, coffee shops, parks, libraries, and leisure centers are to one another. Seek for communities that are walkable and have all the resources needed to encourage healthy living and community involvement.

Evaluate Transportation choices: Evaluate possible neighborhoods' connectivity and transportation choices. Examine the availability of public transit, such as trains, buses, subways, and programs for sharing bikes. Consider how convenient it is to commute to and from major highways and arterial roads, as well as how walkable and bikeable they are. When assessing your options for getting about, take into account things like parking availability, traffic congestion, and transit infrastructure.

Investigate the Housing Market: Examine the housing possibilities, affordability, and market trends in prospective neighborhoods. Find out about the housing inventory, rental rates, median home prices, and rates of property appreciation. When assessing the housing market, take into account variables including housing amenities, housing type, and affordability.

Assess Neighborhood Characteristics: Consider prospective neighborhoods' characteristics and atmosphere. Explore the neighborhood on foot or by car to take in the views of the buildings, greenery, and general atmosphere. Keep an eye out for indications of neighborhood maintenance, community pride, and active participation. When assessing neighborhood character, take into account elements like the neighborhood's aesthetics, historical charm, and sense of community.

Evaluate Community Services: Determine which community resources and services are available in possible neighborhoods. Examine the accessibility of medical facilities, community centers, libraries, places of worship, and social services. When assessing community services, take into account elements including community assistance programs, educational resources, and healthcare accessibility.

Take Environmental Aspects into Account: Take into account environmental aspects that could affect the standard of living in possible neighborhoods. Considerations include the state of the air, noise levels, greenery, and accessibility to nearby natural landmarks. To enhance health and wellbeing, look for communities with quiet streets, pure air, and easy access to parks and green spaces.

Get Input from Locals: Find out what people in the area, neighbors, and community have to say about their experiences residing in possible neighborhoods. Seek advice, opinions, and first-hand stories regarding the features, conveniences, and dynamics of the area. When assessing the desirability of an area, take locals' viewpoints into account.

You can choose a place to live that fits your values, preferences, and way of life by carefully evaluating the neighborhood and community features of possible areas. Making the greatest decision for your future home requires careful consideration of all pertinent aspects, in-person neighborhood visits, and consultation with locals and community organizations.

Investigating Possible Sites and Compiling Data

Making a list of possible places to live and visiting them to get information is an essential part of the process. Here's a detailed guidance on how to visit possible locations and collect information in an efficient manner:

Investigate Possible Locations: Begin by looking for possible locations online. To learn more about potential neighborhoods, cities, or areas of interest, consult real estate websites, neighborhood guides, city directories, and relocation tools. Look at things like trends in the housing market, the cost of living, job prospects, schools, local amenities, and characteristics of the neighborhood.

Make a Shortlist of Potential sites: Using the information you have gathered from your study, make a list of possible sites that fit your needs and interests. Reduce your search to a reasonable number of places you intend to visit in person.

Arrange Your travels: Make ahead plans for your travels to possible destinations. Plan your visits during the weekdays and weekends to gain an understanding of the various times of the day and traffic patterns. Make a schedule for each visit that includes places to go, neighborhoods to check out, and appointments with locals, landlords, or real estate brokers.

Investigate Neighborhoods: To gain a sense of the area, investigate neighborhoods within possible sites. Take strolls or drives in various neighborhoods to take in the views, landscaping, and general atmosphere. Take note of things like neighborhood character, noise levels, cleanliness, and safety. Explore neighborhood features and atmosphere by visiting parks, stores, eateries, and community centers.

Visit Housing Options: To determine suitability and affordability, visit housing options in possible locales. To see the available housing options, schedule tours of rental properties, open houses, model homes, or apartment buildings. Make a note of things like pricing for buying or renting, amenities, size, layout, and condition. When visiting a property, find out about the terms of the lease, the upkeep guidelines, and the local facilities.

Investigate Your possibilities for Transportation: Look into your possibilities for transportation as well as possible commute routes. Take a test drive on the routes you use to commute to work, school, or other locations to evaluate the length of the trip, the amount of traffic, and the availability of parking. Examine accessible and convenient public transit options, such as trains, buses, subways, and bike-sharing schemes.

CHAPTER FIVE

Collect Information from residents: While on your trips, collect information from neighbors, residents, and community members. Start a dialogue to get information about the neighborhood from locals by posing questions to residents, store owners, and passersby. To gain further insight into the location, enquire about community events, safety concerns, neighborhood dynamics, and amenities.

Take Detailed Notes and Pictures: While visiting possible venues, make sure to take thorough notes and pictures. Keep a record of significant facts, insights, and observations regarding every neighborhood, kind of housing, and facilities.

Note the benefits and drawbacks, your preferences and qualms, and any queries or worries you may have about each place.

Evaluate Your Instincts and Feelings: When visiting possible venues, follow your instincts and feelings. Observe how you feel in every housing option and neighborhood. Take into account elements like comfort level, feeling safe, and general atmosphere. To help you make the best choice for your future residence, follow your gut and intuition.

Examine and Compare Data: Following your travels to possible locations, go over and contrast the data, observations, and pictures you took. Consider housing possibilities, neighborhood features, amenities, transportation options, and general suitability when evaluating each place. When comparing, take your priorities, tastes, and financial constraints into account.

You can obtain important insights and firsthand experiences to help you make an informed decision about where to live by physically visiting suitable sites and acquiring information. To choose the ideal spot to call home, do your homework thoroughly, carefully consider each option, and follow your gut.

Choosing a course of action and moving forward

In order to make an informed decision that is in line with your needs, interests, and goals, selecting where to reside requires careful consideration of a number of aspects and actions. Here's a detailed guide to help you decide and go forward:

Examine Your Standards: Go over the standards and preferences you set at the beginning of the process to make sure they still make sense. Think about things like accessibility, cost, security, schools, facilities, transit, and neighborhood features. Establish the elements that are most essential to you and rank them in order of importance.

Compare Possible sites: Using the standards and priorities you provide, compare possible sites. Go over the data, memos, and pictures you took while visiting each place. Think about the advantages and disadvantages of each site, taking into account housing possibilities, neighborhood features, facilities, and available transit.

Think About Trade-offs: When deciding where to reside, think about any concessions or trade-offs you might have to make. Consider each criterion's weight in relation to your overall priorities and objectives. Based on your unique situation and preferences, decide which variables you are willing to prioritize and which ones you are willing to compromise on.

Consult Others: Ask reliable friends, relatives, or advisors for their opinions and guidance. They can offer insightful and important viewpoints. Talk to others about your options, worries, and decision-making process to get their perspectives and ideas. Make the choice that feels right for you in the end, taking into account the opinions and suggestions of others.

Trust Your Instincts: When making a decision, rely on your gut feelings and intuition. Think about your feelings regarding each possible place to live and housing alternative. In addition to following your instincts, think about things like comfort level, sense of community, and general fit with your interests and lifestyle.

Perform Further Research: Perform whatever further investigation or due diligence required to reach a decision. Examine particular facets of the selected area, such as nearby healthcare facilities, educational institutions, employment prospects, and social activities. Obtain any more data or insights that could influence your choice.

Examine Financial Considerations: Examine the price of housing, living expenditures, taxes, and transportation that are related to the place you have selected. Make sure your choice is in line with your financial objectives and budget, and that you are fully aware of the financial ramifications of it.

Decide: Using the standards, priorities, and factors you have assessed, decide in the end where you will reside. Have faith that you have given careful thought to every pertinent aspect and have chosen wisely in light of the facts and resources at your disposal.

Take Action: Put your decision into practice and carry out your planned actions. Get in touch with landlords, real estate brokers, or property managers to start the housing search if you have decided to buy or rent a house. Make plans for the logistics of your move, such as hiring movers or arranging for transportation, if you are moving to a new location.

Accept the Transition: If you've made the decision to relocate, accept the change and the opportunities that come with it. Have an optimistic outlook and be receptive to new experiences, obstacles, and adventures as you approach the process. Make an effort to fit in with your new neighborhood, make friends, and make your selected place feel like home.

You may choose where to live that fits your requirements, preferences, and aspirations by using the procedures listed below to help you make an informed choice. Have faith in your ability to make wise decisions, and take the initiative to make the required preparations so that you may move into your new house with enthusiasm and confidence.

Summary

Choosing where to reside is an important decision that can have a big impact on your future opportunities, well-being, and quality of life. In order to make an informed and appropriate decision, there are a number of important factors that need to be kept in mind during this process. You can confidently and successfully traverse the decision-making process by following the stages listed in this guide:

Establish Your Priorities: Begin by determining your preferences and priorities, taking into account elements like location, cost, security, schools, facilities, transit, and neighborhood features.

Investigate Possible sites: Investigate possible sites in detail, compiling data on housing choices, neighborhood features, amenities, transit choices, and local resources.

Visit and Get Information: To obtain firsthand knowledge and experiences, physically visit possible venues. Investigate communities, housing alternatives, facilities, and transit choices to learn more about the suitability of each area.

Assess criteria and trade-offs: Assess possible sites in light of your priorities and criteria, taking into account any necessary concessions or trade-offs. Examine the benefits and drawbacks of each place to see which best suits your requirements.

Consult Others: To obtain alternative viewpoints and insights, get counsel and guidance from dependable friends, relatives, or advisers. When making decisions, take into account their suggestions and criticism.

Trust Your Instincts: When making a decision, rely on your gut feelings and intuition. Take into account your feelings on every possible place to live and neighborhood, and follow your intuition.

Examine Financial Considerations: Make sure your choice is in line with your financial objectives and budget by carefully examining the financial factors related to each possible location.

Make a Final Decision and Move on: Decide where you want to reside and move on with your plans. Obtain lodging, plan the practicalities of the move, and enthusiastically and positively welcome the change to your new residence.

Embrace the Transition: Take use of the chances presented by your decision to relocate, and make an effort to fit in and establish a sense of belonging.

Selecting the ideal residence involves thoughtful deliberation, investigation, and introspection. You can discover a place to live that suits your needs, tastes, and goals by using these methods and having faith in your decision-making process. This will ultimately improve your general well-being and quality of life.

THE END

9 798888 395813